James McCosh

Ideas in Nature Overlooked by Dr. Tyndall

Being an Examination of Dr. Tyndall's Belfast Address

James McCosh

Ideas in Nature Overlooked by Dr. Tyndall
Being an Examination of Dr. Tyndall's Belfast Address

ISBN/EAN: 9783337429416

Printed in Europe, USA, Canada, Australia, Japan

Cover: Foto ©berggeist007 / pixelio.de

More available books at **www.hansebooks.com**

OVERLOOKED BY DR. TYNDALL..

BEING AN EXAMINATION OF

DR. TYNDALL'S BELFAST ADDRESS.

BY

JAMES McCOSH, D.D., LL.D.,

PRESIDENT OF PRINCETON COLLEGE.

———oo⋅⊙⋅oo———

NEW YORK:

ROBERT CARTER AND BROTHERS.

1875.

PREFATORY NOTE.

As I was about to sail from Great Britain last autumn, on my return to America, I procured a copy of Dr. Tyndall's Belfast Address. I read it on the deck of the vessel; allowed the yeast to ferment in my mind during the voyage; and, on coming home to Princeton College, I delivered my thoughts (not written out) in an Introductory Lecture to the Class of the History of Philosophy. Abstracts of the Lecture were forwarded by my auditors to a few literary journals, and were thence copied into others; and I thought it advisable to write out fully what I had uttered, and to send it, as I was requested, to a periodical, so deserving of encouragement, the "International Review," where it appeared in the opening number of Vol. II.

I find Dr. Tyndall is sending forth edition after edition of his work in England and in America; and some are anxious, I know, to have by them, for their own use or for circulation, a calm reply, free from all personalities. So I have consented to this paper appearing, with some additions, in a separate form.

Dr. Tyndall's Address is now printed with two Prefaces, in which he professes to reply to his critics. The original Address is clear and plausible; but the Prefaces seem to me to be exceedingly weak, loose, and unsatisfactory. His critics had all advanced objections, less or more valid, to his atomic theory of the universe; and some of them had pointed out flaws in his scholarship. To none of these does he condescend to offer an answer. The burden of his Prefaces is, 'See how ill-used a man I am : a bishop has been raising a wail ; the Presbyterians have denounced me ; and the Romish hierarchy are ready to persecute me.' He must not be allowed to forget that he himself began the attack, and is carrying on his defence in quite as offensive a manner as his opponents ; alleging, charitably, that " the common religion, professed and defended by these different people, is merely the accidental conduit through which they pour their own tempers, lofty or low, courteous or vulgar, mild or ferocious, holy or unholy, as the case may be." Those who criticise him are charged with " deliberate unfairness," or with " a spirit of bitterness which desires, with a fervor inexpressible in words, my eternal ill." I happen to know of some of them, that they are praying for him, in all humility and tenderness, that he and all others who have come under his influence may be kept from all evil, temporal and eternal.

Dr. Tyndall thinks that much good may be done in Ireland by the spread of scientific knowledge, as fitted to lessen the bitterness of ecclesiastical feuds. I agree with him here. But, unfortunately, he has only thrown a new element of trouble into the boiling caldron, and, I fear, thrown back the general study of physics in Ireland. I know what I am saying, from having spent sixteen years in that country, which, as an accomplished statesman, at that time Lord-Lieutenant of Ireland, once remarked to me, is less inclined towards scepticism than either of the other two kingdoms, — I am inclined to add, than any country in the world. The denominations which are too much disposed to war with each other have all combined against Dr. Tyndall, — no, not Dr. Tyndall, but the blank theory which he has expounded. He quotes a maxim of Bacon's, — taken, I may remark, from Plutarch, — " It were better to have no opinion of God at all, than such an opinion as is unworthy of him ; for the one is unbelief, the other is contumely." But Bacon, in the comprehensiveness of his mighty mind, has a counterpart enunciation : " I had rather believe all the fables in the Talmud and the Alcoran, than that this universal frame is without a mind." " They that deny a God destroy man's nobility : for certainly man is of kin to the beasts by his body ; and, if he be not of kin to God by his spirit, he is a base and ignoble creature."

It should be noticed that in this paper, under none of its forms, have I charged Professor Tyndall with being an atheist. It is evident that his convictions or feelings have passed through various phases, and are at present very wavering and uncertain, — *feelings*, rather than convictions founded on evidence. It might have been better in these circumstances if he had allowed the mud to settle, and had his mind clarified, before he discussed such subjects as he has done at Belfast. But, as he did raise all this disturbance, he might have been better employed in these Prefaces in telling us what he does believe than in complaining that he has been misunderstood, and in speaking contemptuously of men who know what they believe. In this paper I have made no inquiry into his personal belief (for which he is responsible to God, and not to me) ; but I have felt myself justified in looking at the statements he has published, and at the consequences to which they lead, logically, and in the faith of those who adopt them.

Princeton College, March, 1875.

IDEAS IN NATURE.

ALL throughout his Belfast Address, Professor
Tyndall defends the right of free thought in
such a manner and spirit as to leave the impression
that he imagines that this right has been denied
him somewhere or by somebody. I have not heard
of any one threatening to deprive the savant of his
title to think on all subjects scientific and unscientific. But there are not a few, scientific as well as
unscientific, who doubt whether he showed delicacy or
even propriety of feeling in opening what professes
to be a purely scientific society with such a speculative paper, the more so as no one was allowed to
reply to him in the Association. We often find that
those who use liberty of speech for themselves, are
least inclined to allow a corresponding right to others.
All that is claimed in this paper is the privilege which
he has employed so freely. I feel perfectly entitled to
review his review ; and, in doing so, I appeal to no
other tribunal than the one he carries us to, — the
laws of the Court of Nature.

Dr. Tyndall is not regarded in Great Britain as
a scientific man of the first order: he is not one
of the few stars of the first magnitude. I am not

aware of any discovery made by him which has opened a new department of nature, or set scientific exploration out in a new direction. But he is thoroughly at home in the domain of Physics, and by his researches has advanced certain departments of it. There are some who, on the principle of *ne sutor ultra crepidam,* wish that he would keep within his own magic circle, where he is powerful, and not venture out of it into the wide region of theosophy, where, with his locks shorn by a Delilah in the fascinating form of a love of notoriety, he is no stronger than other men. He doubts whether the great Newton, trained in mathematics and natural philosophy, was fitted to discuss theological subjects. It is a fact that some great biblical scholars take a different view, and speak of Newton as quite capable, by reason of his profound penetration, his long study and deep reverence, to dive into the depth of divine things. It is doubted whether Dr. Tyndall has the same high qualifications ; and those who feel in this way, regret to find him indulging in the construction of theories as to the origin of things, when they would listen to him with great delight dilating on heat and motion, on glaciers and sounds, — and this when they may not be sure that he has come off any higher than second-best in his controversy with Professor Tait, or that he has given the right explanation of the curious phenomena as to sound which he has lately brought before the Royal Society, and which he refers to regions of the air impervious to sound. He is acknowledged on all hands to be a brilliant experimenter and a fascinating expounder ; and his British Association Address is the clearest enunciation and defence of the views of an important school, — constituting a branch of a mutual admiration society — who are ever quoting each other as infallible authorities, — the other members being Professor Huxley, Mr. Herbert Spencer, Mr. Darwin, and Mr. Bain, and a whole host of inferior men who have assisted the leaders in getting the British Association very much under their management, as also certain portions of the London press, and, it may be added, not a little of the college patronage of the late liberal administration of England. We are cherishing the hope that this address, just because it unfolds so openly what was before let out only in hints and prognostications, may tend to produce a reaction in Great Britain ; as men now see — the veil having been lifted from their eyes — whither they are being led. Within the last two years we have seen what a collapse took place when J. S. Mill's autobiography was published, and all men and women discovered into what a dark cavern his philosophy conducted them, with its startling results as to the obligation of marriage ties and the allowableness of suicide, with its avowed want of assurance in life or hope in death: we see that they are "without hope" who are "without God." It is possible that a like recoil may be effected, when all men are made to know that our world consists simply of an interaction of atoms within a limited sphere of space and time, encompassed with an impenetrable region of darkness.

Dr. Tyndall goes back two thousand two hundred years, and quotes a succession of philosophers favor-able to the atomic theory from that time to the present. His historical sketch is adopted at second-hand, and not from the highest authorities.* Eminent as he is as a scientist (to use a phrase not found in Samuel Johnson, but required by the subdivision of knowledge in our day), there is no proof that he has studied philosophy, or that he is specially a philos-opher: he is certainly not a rigid reasoner, and he overleaps wide gaps in constructing his theories. He quotes lovingly such men as Democritus, Epicurus, Lucretius, Bruno, Gassendi, Hume, and Goethe; but has taken no notice of the views of others, usually

* Blunders, such as are sure to be committed by one not master of the subject, and trusting to secondary authorities, crop out ever and anon. Thus he talks of Empedocles "noticing this gap in the doctrine of Democritus;" whereas every tyro in philosophy knows that Empedocles comes before Democritus. Speaking of the cen-turies lying between Democritus and Lucretius, he makes Pytha-goras then perform "his experiments on the harmonic intervals," as if Pythagoras had not died before Democritus was born. He repre-sents Aristotle as preaching induction without practising it; whereas he did practise induction in his natural history, but certainly did not preach it as Bacon afterwards did. He ascribes, it could be shown, a doctrine to Protagoras, the sophist, which no scholar would attrib-ute to him. A writer (Thomas Davidson), in the October number of the "Journal of Speculative Philosophy," proves that he has not given a thoroughly correct account even of the philosophy of his favorite Democritus; whom he represents as making all the varie-ties of things depend on the varieties of atoms "in number, size, and aggregation," whereas Aristotle, the only original authority on this subject, says that he made them depend on the "figure, aggregation, and position." In the same article it is shown that Dr. Tyndall mis-takes throughout in the few allusions he makes to Aristotle.

reckoned the profoundest thinkers of our world, — except, indeed, to speak of the oppression laid on thought by Plato and Aristotle. I mean to supply the inexcusable omission, and to place alongside of the atomic theory the grand truths unfolded by the great philosophers of ancient and modern times ; and show that their anticipations, often vague and mystical, have been made certain by the certain methods of modern science. When these overlooked agencies are mixed up with the atoms, and made to act with them and counteract them, the result may be a harmonious whole, quite consistent with religion, natural and revealed.

It is a well-known historical fact that, somewhere about 600 B.C., there was a remarkable awakening, over many countries, of reflective, as distinguished from spontaneous, thought. From the beginning, men had observed the works of nature, — the seasons, seeds, plants, animals, and the diurnal and annual movements of the heavenly bodies, and turned them to practical use. But, from the time referred to, there were penetrating minds that were not satisfied with practical or phenomenal knowledge ; but insisted on going beneath the surface, and inquiring into the nature and origin of things. In this age appeared Çâkya Munï, the founder of the comparatively pure but inane system of Boodhism ; Confucius, the great moralist of China ; and, according to some, Zoroaster, the reformer of the Magian religion. But the systems of these men were theosophic or ethical, and do not throw any light on the physical phenomena of the universe ; and so we turn to the rise of the Greek philosophy.

Three great schools appear almost simultaneously The inquiry of each is what is the ἀρχή or principle of all things.   One, the Ionian, whose seat was Miletus or Ephesus, explained nature by elements, commonly by some favorite element: as Thales, by water or moisture ; Anaximenes, by air or ether ; and Heraclitus, an offshoot from the school, by fire.   We have here brought before us the deep truth which modern chemistry is unfolding.   The things which we see are compound, and if we would understand them we must trace them back to their components. Another school, the Pythagorean or Italic, whose seat was Magna Grecia, could not be satisfied with these ever-changing elements, and discovered higher and more permanent principles subordinating them in the orderly forms which things are made to assume, and in the numerical relations running through them ; so that, in fact, things are the copies of numbers.   They delighted to trace, often in a mystic way, the properties of figures and of numbers, and were especially the mathematical school of Greece.   They made the earth revolve round the Hestia, or hearth of the universe, and thus suggested the Copernican theory of the heavens. They saw a universally prevalent order, — Pythagoras heard the music of the spheres ; and they called the heaven, from the earth upward, Cosmos, — a word which has been fondly retained as embodying a great truth.   About the same time arose another school, the Eleatic, which affected to go deeper down into the nature of things, and by pure reason found beneath all apparent mutation an essential Being, which has not come into existence, and which is imperishable. The poem of Parmenides opens with an allegory of the soul longing after truth, drawn on by steeds led by virgins along a road untrodden by men, on the road from darkness to light, and brought to the throne of Dikè, who reveals the unchangeable heart of truth. In this we have an anticipation of the doctrine, that the sum of matter and force cannot be increased or diminished by creature action, but remains for ever the same, thus giving a stability to nature.

A hundred years later, and other profound truths are started by great thinkers. Anaxagoras is of Clazomenæ, but removes to Athens (which is to become the eye of Greece), and is intimate with Pericles. Starting from the Ionic point, he is not satisfied that every thing can be accounted for by elements, and he calls in an intelligence (νοῦς) to arrange (διακοσμεῖν) them. When Socrates heard of Anaxagoras bringing in intelligence, he sent for his books, and was astonished, after finding him arranging all things by reason, employing "air, ether, water, and many other things out of place." But this criticism of Socrates, and a like criticism in the next age by Aristotle, show that neither of these philosophers was able to rise to the same elevated position as Anaxagoras ; who was quite consistent in holding that all things might be disposed by Divine reason, and yet be carried on by physical agents, such as " air, ether, and water." The same philosopher contributed another thought. He represented nature as composed of different things, made up of equal parts (ὁμοιομερῆ) ; thus starting the doctrine of definite proportions, which is the true doctrine of all chemistry, and this whether these proportions are caused by atoms or no, or whether indeed there be such things as atoms. About the same time Empedocles, of Agrigentum in Sicily, fabled as perishing in the flames of Etna which he was desirous of looking into, gave to the world another imperishable thought. He used all the four elements of the older philosophers ; but gave to them loves and hatreds, friendships and enmities, drawing them toward each other, and driving them away from each other. This has culminated in the idea of the attractive and repulsive powers of nature. We may allow Dr. Tyndall to give an account of the atomic theory of Democritus, who belonged to Abdera in Thrace. His tenets are : —

"1. From nothing comes nothing. Nothing that exists can be destroyed. All changes are due to the combination and separation of molecules. 2. Nothing happens by chance. Every occurrence has its cause, from which it follows by necessity. 3. The only existing things are the atoms and empty space ; all else is mere opinion. 4. The atoms are infinite in number, and infinitely various in form ; they strike together, and the lateral motions and whirlings which thus arise are the beginnings of worlds. 5. The varieties of all things depend upon the varieties of their atoms, in number, size, and aggregation. 6. The soul consists of free, smooth, round atoms, like those of fire. These are the most mobile of all. They interpenetrate the whole body, and in their motions the phenomena of life arise. Thus the atoms of Democritus are individually without sensation ; they combine in obedience to mechanical law ; and not only organic forms, but the phenomena of sensation and thought, are also the result of their combination."

Some of these points have not been established. One of them seems to combine utterly incongruous things :

it accounts for sensations and thoughts, for pain and pleasure, for love and hate, for judgment and deduction, for ideas of good and evil, for noble aspirations and high purposes, by atoms smooth and round ; as if there was not a fathomless gap between smoothness and sensation, between roundness and reasoning.

Immediately after this appeared the Sophists, who may have done good in some instances by their professional teaching, for which they deserved their fee : but the charge remains that they were not ·seekers after truth ; and it is a fact that they did not add one great principle to the body of philosophy, while they did much to undermine the whole by maintaining that there is no absolute truth, and that truth is only relative to the man who " troweth." Their chief opponent was Socrates, who formally announced one great truth, which all men had been spontaneously discerning and following, that there are purpose and design in every part of the animal frame : pointing to the eye of man, with its delicate structure, and to its eyelids, which open and close for the protection of the organ ; to the ear, which collects the sounds and keeps them separate ; and to the teeth, which in front are fit for cutting, and behind for grinding. He discovers everywhere a providence, and believed himself guided, not by a daimon, but by a daimonion, a divine influence. His great disciple, Plato, rose to a grander, if not a more important, truth, that there is an Idea which has been in or before the divine mind from all eternity, which is the pattern after which all natural things in heaven and earth are formed, and to the

1*

contemplation of which the soul of man, formed in the image of God, may rise as its highest exercise. One of the interlocutors asks whether this paradigm is to be seen in the dust of the earth, and Socrates, who is expounding the idea, is not able to answer; in modern times the scientific man would place the dust under the microscope, and show in it the most beautiful crystalline forms.

But the philosopher who had the most enlarged comprehension of the deep thoughts embodied in the universe was Aristotle, great as a metaphysician, great as a logician, and great as a naturalist. In his usual manner, he employs for explanation a very familiar example, that of a statue of Hercules in a temple. To the question, What is the cause of this statue? four answers may be given: as to its matter, it is made of marble; as to what produces it, it is the workman with his hammer and chisel; as to its form, it is a representation of Hercules; as to its end, it is to adorn this temple. So, in regard to every natural object, we may seek and find four kinds of causes, — using the term cause in a wider sense than we now do: a material cause, the constituents, say elements or atoms; the efficient cause, the power, divine or creaturely, working in it; the formal cause, the order manifested in it, as in the plant or animal; and the final cause, the end which it, say the eye or hand, is meant to serve. I am sure that Aristotle is right in encouraging us to seek for all these causes or principles in nature, and that they are taking a narrow and unsatisfactory view who are overlooking any one of them. In accounting for all things by atoms, Tyndall has seen only one of them, and that the least elevated, — the material cause; and takes no notice, though he knows that they exist, of the forces which make the atoms play, or of the beautiful forms which they assume, and the beneficent purposes which they serve.

The Stoics delighted to dwell on the unity of the universe, and pointed out its perfect harmony. They had an anticipative view of the doctrine that heat will at last absorb all things into itself, out of which a new world will issue. The atomic theory was adopted from Democritus by the Epicureans, and was wrought into a gorgeous form by the Latin poet Lucretius. Neither Democritus nor Epicurus was a professed atheist; on the contrary, both held that the gods made themselves known to man by images or effluxes from heaven. But Lucretius propounds his theory to deliver men from all belief in the gods and superstitious fears, and represents death as the cessation of existence. It is instructive to observe what a run there is in the present day after Lucretius, both by classicists and physicists. He is declared to be the greatest of the Latin poets, and placed above Virgil and Horace. His arguments and his rich descriptions are quoted, and students have to wade through the mantled pool of his erotics to pluck his flowers. It is curious to notice how a philosophy seeks for and creates a poetry suited to it. The philosophy of Epicurus, so prevalent among the Romans, culminated in "De Rerum Natura;" it has to be added, in the

licentious pictures on the walls of Pompeii and Her-
culaneum.   The philosophy of Locke and Bolingbroke
found appropriate verses in Pope.   The subjective
philosophy of Kant came forth in the grand German
poetry of the beginning of this century.   The physical philosophy of our day has already got a sensuous
poetry in works which will doubtless be followed by
others.   It is because philosophy calls forth such
influences, that it comes to have a sway over national
character.   We can believe with Montesquieu that
the Epicurean philosophy exercised an influence in
deteriorating the character of the Romans, in hasten-
ing their ripeness into rottenness, and determining
their fall ; we can understand this when we look into
these fragments of obscene Epicurean verses which
have come out of the fires of Pompeii to testify against
the inhabitants.   We confess that we have fears of
the results when the new physics come to crystallize
into the creed of the rising generation, and to lead the
literature and inspire the prevailing sentiment of the
age.

Dr. Tyndall has no appreciation of the benefit con-
ferred on science by Christianity in introducing new
and lofty ideas : in showing that there is only one
God, and thus preparing the way for the doctrine that
there is a unity in nature ; in leading men to expect
that there are order and wisdom through all God's
works ; in making the study of nature a duty we owe
to God ; and in giving us exalted views of the soul as
fashioned after the Divine image.   He speaks in dis-
paraging language of the scholastic ages, whose func-

tion it was to preserve, all through the cold winter, those seeds which had been deposited by ancient thought, and which were ready to sprout at the return of spring. He might have spoken with more respect of the mediæval ages, had he reflected that in them more new metals were discovered than in all the Greek and Roman times.

It is an interesting circumstance, that Bacon retained the four causes of Aristotle, and gave to each of them an important place ; allotting material and efficient causes to physics, and formal and final causes to metaphysics, which he places above physics. The grand end of science is to discover, first, axioms, or, as we call them, laws of phenomena, and finally causes and forms. Final and formal causes, at the top of the pyramid, lift us up to God. It has often been said that Bacon set aside final causes. This is an entire mistake. Right or wrong, he gave them no place in physics ; but he allotted them the main place in metaphysics, the highest office of which is to carry us to the Supreme.

Mr. Darwin represents Prof. Huxley as the philosopher of his school ! As if to justify this, the professor has of late years taken Descartes under his special protection, though he does not seem capable of understanding, certainly not of appreciating, the deeper tenets of that greatest of French philosophers. The grand merit of Descartes is that he drew the distinction so definitely between matter and mind, between extension and thought, showing that extension had no capacity to produce thinking. Newton, like Bacon, was favorably inclined to the theory of atoms or molecules, but thought it necessary to call in a God to arrange them and make them work harmoniously. His great rival, the highest of all the German philosophers, Leibnitz, in order to account for the operations of nature, felt it necessary to call in, not only forces, but a pre-established harmony. Two horologes keep the same time, not by influencing each other causally, but because of a set of agencies instituted in each and issuing in the same result. So through all nature there is, says Leibnitz, a set of agencies which do so work that every one thing operates in harmony with every other. It is here, if we do not mistake, that God finds the means of answering prayer, which Dr. Tyndall boldly says cannot bring a return.*

He gives us an imaginary conversation between a disciple of Epicurus and Bishop Butler. Epicurus is fitly represented; but I venture to say that, if Butler were alive, he would give a much weightier defence than has been put into his mouth by the President of the British Association. The grand merit of Butler is that he has found in the very con-

---

* Prayer is a duty; and he who prays believes that he will receive an answer in some way, but may not be able to specify the way. It may turn out that the answer comes by pre-established harmony, or, what is the same thing, by the Divine fore-ordination proclaimed in Scripture. The prayer and its answer may be joined, not by physical nor even by causal connection, but in the counsels of God, who has planned, without at all interfering with free will, that there should be both, and that the answer should be brought about by God's own natural agents formed into laws which "continue this day according to his ordinances, for all are his servants." (Ps. cxix. 91.)

stitution of our nature the conscience, as a law which asserts of itself that it is supreme in the mind, and subject only to the great Lawgiver to whom it points. In the same century the Scotch philosopher, Reid, demonstrated that there were principles in our nature, self-evident and irresistible, from which there is no appeal; and the great German metaphysician, Kant, holds that there are forms of thought which are necessary and universal, and that there is a categorical imperative which guarantees the existence of God, the Good. He who holds firmly by these truths may let men employ the atomic theory as they please, to account for the constitution of the universe.

Two great scientific truths have been established in this century. One is the doctrine of the conservation of energy, which implies that all the physical forces are correlated, and that the sum of force, potential and actual, in the universe, is always one and the same. The men who did most to prepare the way for this doctrine — such as Newton, Davy, Oersted, Herschel, and Faraday — all delighted to see God in his works ; and the living philosopher who was the main agent in discovering it, Dr. Mayer, has a mind filled with the presence of God, and looks on force as the expression of the Divine power. The other great doctrine is that of development, acknowledged as having an extent which was not dreamed of till the researches of Darwin were published. How far evolution is to be carried is a disputed point among naturalists. Darwin seems to have a great antipathy to final cause ; but he has somehow or other convinced

himself that there is a God, and is obliged to call in
three or four germs, or at least one germ, created by
God.   It could easily be shown that the doctrine of
development, properly understood, and kept within
inductive limits, is not inconsistent with final cause ;
for we may discern a plan and a purpose, means and
end, in the way in which plants and animals are
evolved, and in the forms they take, which are evidently not by chance, — if the word has any meaning,
— or by blind atoms, but according to a progression
foreseen from the first, and proceeding in a determined
order.

Professor Tyndall thinks he can account for every
thing by atoms, and he reaches the conclusion that
there is nothing but matter.   "Abandoning all dis-
guise, the confession I feel bound to make before
you is, that I prolong the vision backward across the
boundary of the experimental evidence, and discover
in that matter which we in our ignorance, and not-
withstanding our professed reverence for its Creator,
have hitherto covered with opprobrium, the promise
and potency of every quality of life."   "The doctrine
of evolution derives man in his totality from the interaction of organism and environment through countless ages."   A few years ago Dr. Tyndall, in a Lecture
— now published as an appendix to his Address, —
seemed to use different language, and allowed freely
that we cannot see any nexus between cerebral action
and thought, or discover why a movement of the brain
should lead to mental exercise.   But this was never
intended to mean much ; for Dr. Tyndall would say

that just as little do we know *how* oxygen attracts hydrogen. And so he feels himself entitled to hold that matter, though we cannot say how, may give us all the operations of understanding and will.

He accounts for every thing in our world by atoms. This leads us to inquire what we really know about these atoms of which so much is made. First, we seem to be obliged by a sort of necessity of thought or speech to fall back on some such conception. If every thing we see in the world be composite, and capable of analysis and division, we have to think and talk of something indivisible and undecomposable, which we may call particles, molecules, or atoms. But this necessity in thinking does not imply that there are any such actual existences, any more than the corresponding mathematical ideas about points, lines, surface, show that there is such a thing as position without magnitude, or length without breadth, or a surface without depth. For the evidence of the reality of an atom we must appeal, not to pure thought, but to observation. But, then, no one ever saw an atom or handled an atom ; the microscope has not yet been constructed which can see it, nor the balance which can weigh it.

What proof have we, then, of the existence of such indivisibles? The answer, as I understand, is that we require to posit them to account for the nature, the structure, and the operations of material substances. There is, first, the fact that elementary bodies combine in certain proportions. All, however, that this establishes, as our best chemists acknowledge, is only a doctrine of proportions or equivalents. Dalton and others have tried to account for these proportions, by showing that they arise from atoms having specific weights and shapes. The attempt has not been altogether satisfactory, as in chemical combinations the atom, as determined by the balance, frequently exhibits a wide range of deportment, coming under the head of what chemists call quantivalency or atomicity. Secondly, there are the mathematical figures of crystals, which may be supposed to be built up by regular shaped atoms, just as a house is by bricks. Unfortunately, the same substance, sulphur for example, takes allotropic forms which are incompatible ; that is, cannot proceed from any one simple form of atom. Once more, there is the internal mobility of every material substance, which seems to show the constant action of molecules, or at least of something inconceivably small. Such considerations seem to make it probable that there are very small bodies conducting a great part of the actual operations of nature. But every sage man will admit that what we affirm of atoms is only provisionally true. Science in its present state seems to be waiting for some new Newton, Lavoisier, Dalton, or Mayer to furnish the precise conception and expression for what is loose, floating, and somewhat incongruous. It has to be added that there is an increasing number of savants favorably disposed to the theory of Leibnitz, mathematically expressed by Boscovich, and received, though vaguely apprehended, by the great experimental philosopher, Faraday, that

matter consists merely of centres of force acting all around them according to certain laws, and producing that resistance which we attribute to extended bodies. The difficulty pressing on this theory is, Can it account for the inertia of body ? In these circumstances, how rash, with our present knowledge, to account for the whole formation and state of the universe by things of which we know so little !

It is admitted that, by the finest instrument, we can discover matter only in a molar state, that is, in masses. The smallest possible mass is called a molecule. But we are obliged to *suppose* that this molecule is compound : the molecule of water is composed of oxygen and hydrogen ; we can separate the oxygen and the hydrogen, — we *suppose*, the atom of hydrogen from the atoms of oxygen. We cannot have the atom of either of these elements alone or by itself : we can separate the atom of hydrogen only by its being united with something else. Even when we have pure hydrogen, we take for granted that it is composed of molecules having two or more atoms of hydrogen combined.

Atoms are the smallest possible portions of matter which can enter into a combination. According to the common apprehension, they are hard, impenetrable bodies, with a definite shape, which is unknown, and a power of action, of polar action. The negative end of the one attracts the positive end of the other. They act on other atoms all through space, according to the mass, and on every one atom according to the square of the distance. This is in accordance with the doctrine which the author of this paper has long been maintaining, that all material action consists in the mutual action of two or more bodies on each other, probably in the action on each other of two or more atoms.

By far the clearest and most satisfactory account of molecules which we have seen is in a paper read before the British Association at Bradford, in 1873, by Prof. Clerk Maxwell, of Aberdeen. The mass, weight, and properties of a molecule are unalterable. Though indestructible, it is not hard or rigid, but is capable of internal movements, and when they are repeated it emits rays. They are flying all through the atmosphere, quicker than a cannon-ball, at the rate of about seventeen miles in the minute, and they diffuse throughout nature, matter and momentum and temperature. We know of three distinguished men who have been trying to discover their size and weight: Loschmidt, Mr. Stoney, of Dublin, and Sir William Thomson, of Glasgow ; and, it is calculated that about two millions of molecules of hydrogen in a row would occupy a millimetre ; and that in a cubic centimetre of any gas, at a standing pressure and temperature, there are about nineteen million million million molecules. A million million million million of them would weigh between four and five grammes.

Mr. Maxwell arrives at a much more philosophical conclusion than Dr. Tyndall : " The exact quality of each molecule to all others of the same kind gives it, as Sir John Herschel has well said, the essential character of a manufactured article, and precludes the idea of its being eternal and self-existent." He discovers in the

very nature and properties of a molecule a proof of
design : " A 'collocation,' to use the expression of Dr.
Chalmers, ' of things which we have no difficulty in
imagining to have been arranged otherwise.' " He
thus closes: " Though in the course of ages catas-
trophes have occurred and may yet occur in the
heavens, though ancient systems may be dissolved and
new systems evolved out of their ruins, the moleclues
out of which these systems are built — the foundation-
stones of the material universe — remain unbroken and
unworn. They continue this day as they were created,
perfect in number and measure and weight ; and, from
the ineffaceable characters inpressed on them, we may
learn that those aspirations after accuracy in meas-
urement, truth in statement, and justice in action,
which we reckon our noblest attributes as men, are
ours because they are essential constituents of the
image of Him, who, in the beginning, created not only
the heaven and the earth, but the materials of which
heaven and earth consist."

To show how little agreement there is among scientific men as to the constitution of matter, I may
quote the language of an original observer and a sug-
gestive writer, Prof. T. Sterry Hunt, in a paper read
at the Centennial of Chemistry : " In chemical change
the uniting bodies come to occupy the same space at
the same time, and the impenetrability of matter is
seen to be no longer a fact, the volume of the combining masses is confounded, and all the physical
and physiological characters which are our guides in
the region of physics fail us, gravity alone excepted :

the diamond dissolves in oxygen gas, and the identity
of chlorine and sodium are lost in that of sea salt.
To say that chemical union is in its essence identifi-
cation, as Hegel has defined it, appears to me the
simplest statement conceivable.   The type of the
chemical process is found in whatever form which it
is possible, under changed physical conditions, to
regenerate the original species.   Can our science
affirm more than this ? and are we not going beyond
the limits of a sound philosophy, when we endeavor, by
hypotheses of hard particles with void spaces, of atoms
and molecules with bonds and links, to explain chemical affinities ?   And when we give a concrete form to
our mechanical conceptions of the great laws of definite and multiple proportions to which the chemical
process is subordinated, let us not confound the
image with the thing itself ; until, in the language of
Brodie, in the discussion of this very question, we
mistake the suggestions of fancy for the reality of
nature and we cease to distinguish between conjecture
and fact."   The difficulty is not removed by this doctrine, nor is the subject made more comprehensible by
the Hegelian expression (for it is nothing else) identi-
fication ; for we still put the question, what are the
things which thus occupy the same space at the same
time, which are thus dissolved, thus seen to be identi-
cal ?   Atoms ? or what else ?

Atoms and molecules are admissible, because they so
far account for the shapes and activities of molar matter falling under the senses.   But they do not explain,
and do not even seem to explain, the laws and opera-

tions of mind, — of sensation, judgment, reason ; of love, passion, resolution. There is no proof that there is sensation in any one of these atoms, or that sensation will be produced by two or more of them striking against each other. We may be able to account for the shapes of a stone or mountain, of a planet or star, by atomic agglomerations. But can we, with any appearance of plausibility, account in this way for the affection of a mother for her son, of a patriot for his country, of a Christian for his Saviour ? Aggregate them as you choose, and let them dance as they will, there does not seem to be any power in them to generate the fancies of Shakespeare, — his Hamlet, his Lady Macbeth, his King Lear, — the sublimities of Milton, the penetration of Newton, or the moral grandeur of the death of Socrates. We can conceive them to fashion the bodily shape of Prof. Tyndall as he addressed the Belfast audience ; but we have some difficulty in conceiving how they should compose the discourse which he delivered, — not only the words but the thoughts, the theories, — and give rise to the approbations and disapprobations in the minds of his audience. Atoms may come in appropriately enough in the one case ; but all, except those who have gazed so long on them that they have become magnified beyond their proper bulk, feel that they have no fitting place in the other. What — to employ the very mildest form of rebuke — can be the use of devising hypotheses which have not even the semblance of explaining the phenomena ? In the interest of science, not to speak of religion, it is of moment at this

present time to lay an arrest on such rash speculations ;
and to insist on scientific men refraining from what
Bacon denounces as "anticipations of nature," and
confining themselves to facts and the co-ordination
of facts.

"I am blamed," says our lecturer, "for crossing the
boundary of the experimental evidence. I reply that
this is the habitual action of the scientific mind, at
least of that portion of it which applies itself to physical investigation." He tells us that "the kingdom of
science cometh not by observation and experiment
alone, but is completed by fixing the roots of observation and experiment in a region inaccessible to both,
and in dealing with which we are forced to fall back
upon the picturing power of the mind." Is this a safe
ground on which, it seems, a certain portion of the
scientific mind has fallen back, "upon the picturing
power of the mind"? Every one knows how apt the
mind is to picture things which have no reality, and
how apt every mind would be, on such a system, to
draw its own pictures. It is surely time to lay a re-
straint of a stringent kind upon the use, or rather
abuse, of the imagination in science. It is curious to
notice that, while M. Comte, the founder of the Positive School, sought to restrain science to the observation of phenomena, meaning by phenomena only
sensible phenomena, the school which has sprung from
him has broken his trammels, and is revelling in all
sorts of hypotheses, about the origin of things, about
world-making and world-ending. Mr. Mill is partly
responsible for this. The book on Induction in his

"Logic" is a very valuable one ; but he has dwelt more on the mental processes involved than on what Bacon places first and last, — the gathering of facts, the collation of facts, with the "necessary rejections and exclusions." The process recommended by Mill is deduction rather than induction: it consists in forming hypotheses, in deducing from them, and in verifying them. So we have now, cold Positivism having been broken up, a freshet of hypotheses: the atomic hypothesis, the development hypothesis, the pangenesis hypothesis, — no one of which, it is acknowledged, is capable of direct proof. I am not maintaining that hypotheses should never be devised. But there never was more need than now of imposing restrictions upon them. First, an hypothesis should not be started till there has been an extensive induction of facts ; and the hypothesis should grow out of the facts, and not out of the picturing power of the mind. Secondly, it should be regarded by those who advance it, and be announced by those who use it, as a mere hypothesis, till such time as it is established. Thirdly, the apparent exceptions should be noted and stated ; *i. e.*, the hypothesis should be modified so as to take in these, and not be adopted till it explains them. Fourthly, an hypothesis, as long as it is a mere hypothesis, should not be employed to establish a doctrine, say a religious or an anti-religious one.

I admit that we may legitimately cross the boundary of experimental evidence, taking "experimental" in its restricted sense. But as we do so, let us know and acknowledge that we are doing so, and clearly announce what other method we are following, and let not this be the pictorial one. Physical science, as science, should be rigidly confined to experimental evidence ; and, as the " British Association " professes to be for " the advancement of science," the places inviting the meetings should let those who manage the society know that they should confine themselves to their own rich and ample domain.

I acknowledge that there are means of reaching truth other than mere experiment. Mental, as well as material facts, are to be observed and weighed by those who would reach the higher and deeper verities of nature. Some truths are known by intuition, and called first truths ; some are reached by deduction, as in mathematics ; and more by a judicious combination of intuition, induction, and deduction. But let these methods, and the principles or facts they employ, be distinguished in the mind of the investigator, and be kept separate in the exposition of his views. The mixing of these things leads to their being confounded, and the issue is utter confusion. How profound the wisdom in the warning of Bacon, " This folly is the more to be prevented and restrained, because not only fantastical philosophy, but heretical religion, spring from the absurd mixture of things divine and human ! " Bacon maintained that men could go beyond mere material and efficient causes to other and higher, — to final and formal. But he allotted these last to a separate department. While he confined physics to material and efficient causes, he reserved for metaphysics the inquiry into the final and formal.

Combining the other allowable modes of inquiry with the experimental, we may discover great principles overlooked by Tyndall, but having a deep foundation in nature. Let us look at some of these.

Intelligence. Dr. Tyndall refers to some great man not named by him. " Did I not believe," said a great man to me once, " that an Intelligence is at the heart of things, my life on earth would be intolerable." Surely Dr. Tyndall's acquaintanceship must be confined to a very small circle, if he has only met with one man uttering such a sentiment. It is the spontaneous feeling of humanity. Anaxagoras only expressed what all men, not led astray by sophistry, had felt ; and he was farther right when he believed that the presence of Intelligence was quite compatible with the operation of physical agents.

We are not inquiring at present whether pantheism or theism is the right view, whether the intelligence is in nature or beyond nature : this subject will be taken up farther on. We are not inquiring whether there is an inherent life in nature, or whether its activity springs from exquisitely nice adaptations made by a power above them. In either case we are compelled, if we would account for, if we would get a solution of, what is evident, to maintain that there is mind in nature. Professor Tyndall gives a clear account of the Lucretian way of explaining apparent design : " The interaction of the atoms throughout infinite time rendered all manner of combinations possible. Of these, the fit ones persisted, while the unfit ones disappeared. Not after sage deliberation did the atoms station themselves in their right places, nor did they bargain what motions they should assume. From all eternity they have been driven together; and, after trying motions and unions of every kind, they fell at length into the arrangements out of which this system of things has been formed." Bacon and Newton were favorably inclined toward the atomic theory of matter; but then they thought that blind atoms were as capable of working disorderly as orderly, of producing evil as producing good, and in the order and benevolence in the world they saw proofs of an organizing power. " Even that school," says Bacon, " which is most accused of atheism, doth the most demonstrate religion; that is the school of Leucippus and Democritus and Epicurus. For it is a thousand times more credible that four mutable elements and one immutable fifth essence, duly and eternally placed, need no God, than that an army of infinite small portions or seeds, unplaced, should have produced this order and beauty without a divine marshal."

" But it is said that the fit survive while the unfit perish. We are inclined to discover an ordinance of intelligence and benevolence in the very circumstance that there is a fitness, and that the fit survive. Things might all have been such that there was no fitness in them, and the most unfit might have survived. That things are otherwise, we can explain only by supposing that in the original structure of the atoms there was a fitness to produce fit things, and to secure that they should survive. We hold that the forms or potencies, one or both, of atoms must originally have been such as to make them fit for building up the temple. The fit survive because they have the fitness to do so, and are placed in a state of things in which they can survive because of their nature. It is conceivable that things might have been otherwise, that the atoms might have been such as to be incapable of order, and the unfit have survived to work never-ceasing disorder; and, when sentient beings appeared, to produce only misery. But it is said that in that case the suffering would instantly perish. Yes, as things are now constituted; but things might have been so constituted that the suffering could not perish, that innocent sufferers must suffer for ever. All those assumptions about the fittest surviving proceed tacitly on the principle that there is an established fitness in things so to do.

Final cause. On this point Socrates was only expressing what all thinking minds have spontaneously felt from the beginning, that there is evident purpose in the universe; means and end, — the means being also ends and the ends means to something farther. Aristotle placed the whole subject in a truly philosophic position, when he showed that we should seek for four kinds of explanatory causes or principles in nature. We may seek for a material and efficient cause : these are the atoms, and the forces for such there must be — operating in them. But, then, we may also seek for a formal and final cause: in the atoms being made by their forces to assume the shapes which we see in the plant and in the animal, and to conspire to fashion the ear by which we hear, and the eye by which we see, and the hand by which we grasp. There is no inconsistency, though narrow minds may be led to believe that there is, between these different kinds of causes. The matter of the universe and the powers of the universe are made to combine and conspire to produce these beautiful laws and types, and accomplish these beneficent ends. The discovery of efficient cause does not set aside final cause. The final cause is in many cases more obvious than the efficient. That the coats, humors, shapes, and nerves of the eye were made to combine to form an image on the retina whereby the percipient sees, is a proof of intention, and this whether physiologists are or are not able to discover the processes by which the eye is produced.

It is a characteristic of the whole school of materialists that they speak disparagingly of final cause. And I confess at once that some defenders of natural religion at times speak of God as if he were a mere mechanician, — a sort of higher mechanist, or clockmaker. I farther allow that there are minds which dwell only on curious fitnesses and small providences, and in fact discover in nature purposes which God never intended. We, whose range of vision is so limited, should conduct our inquiries into the intents of an omniscient God, with humility and the profoundest reverence. By all means let us notice those nice adaptations and minute providences everywhere forcing themselves on our attention, but let us so widen our vision as to see that these are fittings of a very large machine or organism, in which the ends are means and the means are ends, and in which the particular providences are essential parts of a universal providence which looks to the whole, and makes every part conspire to the good of the all.

Hugh Miller, in criticising "The Vestiges of Creation," remarks that there is nothing in the doctrine there set forth inconsistent with final cause or the belief in the existence of God, though it seems to be incompatible with the Scripture account of the origin of man. Agassiz and others have shown that there is a plan in the way in which plants and animals have appeared on the earth, and the evidence of this would not be set aside though we should discover that this was produced by natural selection, or some other physical agency. Even though the Darwinian theory should turn out to be true in all its main principles, as it is certainly true in some of its principles, there would still be traces of design everywhere in nature in the manner in which natural agencies have been made to conspire to produce beneficent ends. I am convinced that when the method of God's procedure in producing animated beings is fully unfolded, it will display innumerable traces of the fitness of the time and way in which new species have been introduced, whether by natural or supernatural means. But the advocates of this theory, led by Mr. Darwin himself, have, commonly, been speaking contemptuously of final cause, and been seeking to efface all the inscriptions on nature which seem to read, "I am a creature of God." Yet in spite of their opposition to teleology, these men are coming face to face with

striking examples of it.   Dr. Tyndall, gives us one of
these from Mr. Darwin : " Take the marvellous observation which he cites from Dr. Crüger, where a bucket,
with an aperture serving as a spout, is formed in an
orchid.   Bees visit the flower ; in eager search after
material for their combs they push each other into
the bucket, the drenched ones escaping from their
involuntary bath by the spout.   Here they rub their
backs against the viscid stigma of the flower, and
obtain glue ; then against the pollen masses, which
are thus stuck to the back of the bee, and carried
away."   He then quotes Darwin : " ' When the bee,
thus provided, flies to another flower, or to the same
flower a second time, and is pushed by its comrades
into the bucket, and then crawls out by the passage,
the pollen mass upon its back necessarily comes first
into contact with the viscid stigma,' which takes up
the pollen ; and this is how that orchid is fertilized.
Or, take this other case of the *Catasetum.*   'Bees
visit these flowers in order to gnaw the labellum ; in
doing this they inevitably touch a long, tapering, sen-
sitive projection.   This, when touched, transmits a
sensation of vibration to a certain membrane, which
is instantly ruptured, setting free a spring, by which
the pollen mass is shot forth like an arrow in the
right direction, and adheres, by its viscid extremity,
to the back of the bee.'   In this way the fertilizing
pollen is spread abroad."

Tyndall tells us that Darwin's books are a " reposi-
tory of the most startling facts of this description,"
as, for instance, his account of the ways in which

insects and birds carry the pollen from one plant to another. In due time a Paley will arise to furnish proofs of design from such facts as these. Darwin will supply the facts, and we are just as capable as he of perceiving their meaning. He may reject teleology, but his facts are teleological whether he acknowledges it or no.

Professor Huxley has a good deal of the Arab in his character, and rather delights to have his hand against every man — except those of his own tribe ; but is irritated, I am told, when he finds, in consequence, every man's hand against him. His Bedouin attacks show courage, and make him a favorite with John Bull, who likes openness of speech. There is also, I suspect, some irony in his nature. He must have been in rather a quizzing humor, when he discussed, before a Belfast audience, the Cartesian question, whether the lower animals are mere automata, and urged so many arguments to show that they are, adding that these arguments had not convinced him. My idea of his secret intention in this lecture is, that he means to drive us to some sort of potential life, or pangenesis in all matter. In conducting this discussion, he furnishes us with a very beautiful instance of adaptation in the animal frame, an adaptation altogether independent of the mind or will of the animal. He takes a frog deprived of senses and of feeling, and he puts it on his hand : " If you incline your hand, doing it very gently and slowly, so that the frog would naturally tend to slip off, you feel the creature's forepaws getting a little on to the edge of your hand until he can just hold himself there, so that he does not fall ; then, if you turn your hand, he mounts up with great care and deliberation, putting one leg in front and then another, till he balances himself with perfect precision upon the edge of your hand ; then, if you turn your hand over, he goes through the opposite set of operations, until he comes to sit with perfect security on the back of your hand. The doing of all this requires a delicacy of co-ordination, and an adjustment of the muscular apparatus of the body, which is only comparable to that of a rope-dancer among ourselves."

We are glad to have the description of the fact from Mr. Huxley ; and we reckon ourselves quite as entitled to judge of its meaning as he is. But they tell us that we are not to look on this wonderful adjustment as implying design or a purpose : this is degrading to God, as making him a mere artificer, and is a technic, mechanical, anthropomorphic view. Now, it is always to be borne in mind, that God is represented as saying, "My thoughts are not your thoughts, neither are your ways my ways, saith the Lord." The error of anthropomorphism, of which the school have such a horror, does not consist in supposing that God has qualities like those of man. But it consists in holding that God has no other qualities but those which man has, or in maintaining that these exist in God after the same manner as they do in man, or in attributing to the Divine Being the weaknesses of humanity. We shall have to abnegate our intelligence, if we are not allowed to discover an intelligence in nature as we discover intelligence in human workmanship. We are not degrading God when we ascribe to him the wisdom which we see exhibited in a small way by his creatures, provided we make it infinite in extent. We do not impose our qualities on the Divine Being ; but we claim to be formed in his image, and to reflect something of the light of his perfections.

Laws and Types. This was the grand truth expounded by Plato under the name of Ideas, and carried out by Aristotle under the designation of Formal Causes. Every one sees it in the seasons and revolutions of the heavenly bodies, in the plant, in the animal, and the human form. All science illustrates it. The laws of physics and chemistry are expressed in numbers implying definite proportions. The guiding principle in botany and zoology is type ; that is, regulated structure and model form. The laws of nature, as they are called, are most of them complex, being the result of arrangements with conspiring agencies. This is the case with the seasons, with the elliptic orbits of the planets, with the cycles of the sidereal movements : all are constructions in which various matters and forces combine and co-operate. Possibly all these constructions may carry us back ultimately to the forms and properties of atoms and their collocations ; but in that case there must have been a plan in what has produced such results.

A Universal Harmony. The Pythagoreans sought for a music in all nature. The Stoics maintained that the harmony was perfect, and ascribed it to the Fatum, — the word or will of Deity. Modern science establishes what was then a mere surmise. Astronomy shows us order and uniformity in the utmost regions of space. Geology exhibits the same laws operating for unnumbered ages. The spectroscope discovers the same elements in the most distant stars as we have on our earth. The doctrine of the conservation of force lets us see how it is that our world is so stable, while it points not unobscurely to a time when all things will be burned up.

This harmony appears to the writer of this paper to take two forms. First, there is the adaptation of the properties of one body to those of another, whereby they act and react on each other, atom on atom, molecule on molecule, mass on mass, all to produce harmonious results. Secondly, there is the pre-established harmony propounded by Leibnitz, — a harmony produced not by one body acting on another, but by the original disposition of agents, whereby results are produced which fit into each other.

Life. The whole school are obliged to confess that they cannot explain every thing by atoms or by any machinery at their disposal. They acknowledge that there is no known law of nature which can bring animated beings out of inanimate objects. Dr. Tyndall indeed says : " Those who have occupied themselves with the beautiful experiments of Plateau will remember that, when two spherules of olive oil, suspended in a mixture of alcohol and water of the same density as the oil, are brought together, they do not immediately unite. Something like a pellicle appears to be formed around the drops, the rupture of which is immediately followed by the coalescence of the globules into one. There are organisms whose vital actions are almost as purely physical as that of these drops of oil." True, but these drops of oil are after all physical and not organic. Mr. Darwin, to help him out of his difficulties, is obliged to call in more than natural selection. He holds that there is a pangenesis or panzoism in all animated being. Now, what is this but the "life" of the old zoologists whom they so ridicule? It is clear that, after they have made atoms perform all sorts of dances, there still remains a residuum which atoms cannot explain ; and it would be wiser in them, before they go on speculating so wildly, to employ years of patient inductive observation and experiment to determine what this — I will not call it life, but — pangenesis is.

Mr. Darwin is obliged, to account for life, to call in three or four original germs, or at least one germ, created by God. Dr. Tyndall and the younger members of the school are not satisfied with this compromise. "The anthropomorphism which it seemed his object to set aside is as firmly associated with the creation of a few forms as with the creation of a multitude." Not satisfied with Darwin, he falls back on Spencer. Mr. Spencer has given us one of the weakest and most unsatisfactory definitions of life ever propounded : "It is a continuous adjustment of internal relations to external relations." Dr. Tyndall says : "The organism is played upon by the environment, and is modified to meet the requirements of the environment." The difficulty is dexterously avoided by this loose statement. For the difficulty is to get the organism which is to act on the environment. It is the action of the organism on the environment, the action of a living body on inanimate matter, that is the thing to be accounted for; and this is carefully avoided.

Mind in Man. It is at this point that the theory is felt to be weakest — is seen visibly to break down. There is no appearance of plausibility in the statement that atoms can produce sensation, pleasure or pain, sense-perception, memory, judgment, desire, or will. Viewed *a priori* the two ideas seem to be of an entirely different order, extended matter and thought. Experience furnishes no example of mental affection produced by bodily action. They hint, indeed, and would like to tell us that thought may have existed in the atoms from the very first, if not actually, at least potentially. And then, in carrying out their theory, they are obliged to admit that for millions of millions of years this thought, all along in the atom, did not come forth in any actual thinking. We could believe all this if we had evidence; but even then we would insist that when, at the end of these countless years, thought came into exercise, it must be by some power calling it forth, and this power must itself be a thinking power.

But then it is said that by this theory we are merely exalting matter, and not degrading mind. Dr. Tyndall tells us that he remembers the time when " I regarded my body as a weed." There has been a terrible reaction of opinion since that time. It is possible that the one extreme may be as far from the truth as the other. The Scriptures represent the body and soul as the two essential constituents of man's compound nature. They would have us cherish both, and believe that the two are to be reunited at the resurrection. It is admitted that, in the ordinary state of matter, — the state of air, water, metal, earth, bone, muscle, skull, — it has no capacity of thought or feeling. But then it is supposed that in some refined form — say as nervous structure — it may rise to intelligence and feeling. He has, however, to allow in his Appendix, " Granted that a definite thought and a definite molecular action in the brain occur simultaneously, we do not possess the intellectual organ, nor apparently any rudiment of the organ, which would enable us to pass by a process of reasoning from the one to the other." He speaks of the chasm between the two classes of phenomena being "intellectually impassable." If this be so, the attempt to resolve mind into matter has no plausibility whatever.

When we press the school with the first truths of Aristotle, the intuitive principles of Reid, and Kant's forms of sense, understanding, and reason, they fall back on Herbert Spencer's boasted resolution of them. David Hume and J. S. Mill accounted for these — that is, for our belief in such truths as mathematical axioms, the existence and identity of self, and the universality of cause and effect — by association of ideas. As the author of this work has labored hard to disprove this theory, he is glad to find it abandoned as utterly unfit to explain the nature of truths claiming the necessary consent of all men. Herbert Spencer defends a universal postulate, that we must assume a proposition of which we cannot conceive or think the opposite. (" Principles of Psychology," Chap. XI.)* Using this as a test or criterion, he has a whole host of *a priori* truths which he does not attempt to enumerate or to classify, as mathematical axioms and arithmetical propositions, an objective existence external to consciousness, and an Unknown Reality hidden under all the shapes that present themselves. Indeed, he is threatening, to the astonishment and dismay of scientists, to find an *a priori* demonstration of physical laws which are usually supposed to have been discovered by induction.† Mr. Mill accounted for these by associations formed in the experience of the individual; but Mr. Spencer is lauded because he has constructed a much more comprehensive theory. He calls in the experience of the race, including all our animal forefathers, from the ascidians downwards. " Instead of relatively feeble nervous associations

---

* Following Hamilton, who follows Kant and Leibnitz, he makes the primary mark of first truths to be Necessity. I have endeavored to show that it is Self-Evidence. We look on things and the relation of things, and discover them to be so and so. As we do so, we cannot be made to think the opposite; and this becomes the secondary test, which is Necessity. As all men perceive in the same way, we have a third criterion, Universality This makes ultimate truth to consist, not in the associations of the individual (with Mill), or of the race (with Spencer), or of necessity of conviction (with Leibnitz, Kant, and Hamilton), but in the perception of objects.

† An able writer, in the " British Quarterly," is greatly perplexing him in regard to this point.

caused by repetition in one generation, we have organized nervous connections caused by habit in thousands of generations, — nay probably millions of generations. Space relations have been the same, not only for all ancestral men, all ancestral primates, all ancestral orders of mammalia, but for all simple orders of creatures. These constant space relations are expressed in definite nervous structures, congenitally framed to act in definite ways, and incapable of acting in any other ways. Hence the inconceivableness of the negation of a mathematical axiom, resulting as it does from the impossibility of inverting the actions of the correlative nervous structures, really stands for the infinity of experiences that have developed these structures." ("Psychology," Chap. XI.) I venture to predict that the boasted discovery of Spencer will not run so long a career as the association theory of Hume and Mill ; and that it will be seen in the end to be as incapable of accounting for the phenomenon of all men perceiving certain truths intuitively, and being incapable of thinking the opposite.

For, observe that the accumulation of the experiences of the individual in Mr. Mill's theory is mental throughout, and is in a sense intelligent, aided by associations in consciousness. Mr. Spencer's experiences consist in the formation of "definite nervous structures." How consciousness and intelligence ever get into these structures he does not tell us, and does not profess to tell us. But "hence the inconceivableness of the negation of a mathematical axiom." This is a fair specimen of his agility in leaping over lacunæ in reasoning without his calling our attention to them, or even, so far as we can judge, seeing them himself. What a gap between a nervous *structure* and the *inconceivableness* of a falsehood! I believe Mr. Spencer would allow that the connection is unthinkable. Yet it is by means of these unthinkable bands that he builds his theory.

I admit the existence of hereditary tendencies. They are very much the result of bodily organization; such as the aptitude of dogs to point to game or assist the shepherd in guarding his flock, or the disposition in men and women towards certain movements and appetites. They are produced originally by circumstances, are continued by habit, and fashion the brain structure, which may become hereditary. But these surely are different in their whole nature from those fundamental perceptions and convictions which are in the very structure of our minds, which gaze immediately on things and on truth, and carry with them their own validity; as that two and two make four, that every effect has a cause, and that there is an essential distinction between good and evil. These are in all men, and in no brutes: can any one bring himself to believe that they are in the primates, or the ancestral orders of mammalia, — not to go back to the ascidians? In a nascent state they are in infants and savages, and come forth in adults, and can be expressed by educated minds; but cannot be developed in the souls of lower creatures. They look at truth self-evident, and carry with them a necessity of conviction. I should like to dwell on this topic, as it is the ground on which the whole school fall back. An insecure ground it is at best; and is known, acknowledged, and felt to be so, — the ultimate foundation of truth is not things perceived, but an aggregation of human experiences flowing from and determined by a succession of anterior, unintelligent, animal experiences. But I have said enough to counteract the assumption of Dr. Tyndall, that in the end truth *rests* (the word is a mockery) on the flowing stream of " infinitely numerous experiences received during the evolution of life." Spencer has confounded two things which ought to be carefully separated: a propensity or tendency to feel and act in a particular way, and capable of becoming hereditary, with a principle of reason which has been in man's nature from the beginning, and gazes on and guarantees immutable truth.

A personal God. This is the result of the separate truths which have passed before us. The traces of intelligence, of purpose, of order, of harmony, of life, of thought in man, who is conscious of personality, all carry us up to One who is the cause, and who must himself possess the qualities which he has produced.

Dr. Tyndall does not wish to be called an atheist. In commenting on a resolution passed by the presbytery of Belfast, he declares that he merely ignores the existence of *their* God. But what are the nature and character of the God retained by him? It is a God unknown and unknowable, as Tyndall expresses it, — " a power absolutely inscrutable to the intelligence of man." In this style of remark the materialists are led

by Herbert Spencer, who took advantage of, and followed out to their consequences, certain rash expres-
sions employed by Sir W. Hamilton and Mansel, —
the two leading philosophic authorities at the time
when this modern Titan was commencing his war
against the gods who rule in Olympus. Mr. Spencer
condescendingly hands over this unknown land to
religion, which, however, has shown no inclination to
part with its rich inheritance in possession for a title
to a property in the *terra incognita.* In that Book
from which so many take their religion, God is represented as so far unknown, because we are finite and
he is infinite, but also so far known because we are
formed in his image. " The heavens declare the glory
of God."   " The earth is full of his praise."   Paul
did observe in Athens an altar with an inscription to
the unknown God ; but he takes advantage of this to
say, " Whom therefore ye ignorantly worship, Him
declare I unto you."   And he declares that the invisi-
ble things of God are clearly seen, " being understood
(νοούμενα, the strongest word the Greek language can
supply) from the things that are made, even his eternal power and godhead."   The inspired writers everywhere encourage us to seek and to know the Lord.
What a miserable prospect, to be obliged to look out
for ever on this impenetrable darkness, where there
may indeed be a power, to which, however, we would
feel as little inclination to pray as to a cold mountain
or the hard rock. Surely they are " miserable com-
forters," who have nothing to say, when they are
brought, as they often must be, into the presence of

the widow, the fatherless, the motherless, of those suffering from incurable disease, the despairing and the dying, except " There is a mysterious power beyond the visible ; but ye need not look to it, for you cannot know whether it has any love or pity for you."

I make no inquiry into the personal belief of Dr. Tyndall. But I am entitled to examine and criticise the statements in regard to God which he has so ostentatiously made. We see what he condemns and rejects ; we are not so sure about what he believes. The great body of theists think that they have proof of the existence of a God as the cause of nature, above nature, independent of nature, which He has created and continues to preserve. Our author evidently sets aside this view. He is not willing to allow us a God outside of nature. He is obliged, however, to admit a "formative power, as Fichte would call it, this structural energy ready to come into play and build the ultimate particles of matter into definite shapes" (Appendix). This might seem to make him, like Fichte, a pantheist : but he is not inclined to become fixed down to any religious creed ; and, so far as I can see, retains nothing of pantheism but its sentimentality, to which so many are clinging, — wishing to keep alive the flower after they have cut down the tree on which it grew. Whatever this God may be, he must be material if all things can be derived from matter. But, in fact, the school is not entitled to say any thing about their God, for he is and must be unknown.

The question may be put, and is put, What evidence have we that there is such an unknown power? On their principles, they have none. They tell you that there is a necessary conviction which requires a belief in something beyond the visible. But the question arises, May not this necessary belief be accounted for in the way in which Mr. Spencer accounts for other necessary beliefs, by ascribing it to an hereditary feeling, gendered by our ever coming to something unknown? Whatever the fathers of this nescient philosophy may do, from some remaining hereditary feeling handed down from the superstitious age of their ancestors, and not yet obliterated, the children trained by them are marching on in the road which has been opened to them, and affirming that we have no reason whatever for believing in this unknown region, except a subjective feeling which we can account for, and which will disappear in a few generations. This young race of thinkers will farther tell you, and others will agree with them, that this unknowable God is not worth contending for.

In order to furnish some sort of satisfaction to themselves when they feel how little they have left, and not to scare others by the emptiness and loneliness of the prospect, materialists are ever falling back on some *unknown power.* But if they know it to be a *power*, they know something of it : it is not absolutely " inscrutable." We ask them how they know it to be power, and we show them that on the same grounds we may know it to be something more, — to be vastly more, to be also intelligence, wisdom, and goodness. Every one who has thought on the subject perceives how large a portion of our knowledge is obtained by the use of the principle of cause and effect. It is a favorite maxim of Aristotle that we can, properly speaking, be said to know things only when we know their causes. How do we reach such a common truth as that the persons walking past us on the street are beings possessed of intelligence and feeling? It is evident, on the one hand, that we do not by the senses perceive their souls as we perceive their bodies ; and, on the other hand, that we are not immediately conscious of their souls as we are of our own. We are certain that we are surrounded by intelligent men and women, because we see effects which we know from our own experience imply an intelligent cause. It is on a like principle that we argue from the visible effects in the world that there is a power beyond, — a power so far known. But by a like process, that is by the argument from effect, we argue that there must be a benevolence beyond, to account for the benevolence we see in nature.

Prof. Tyndall tells us, in a passage of his first Preface, in which he seems to express genuine feeling : " I have noticed, during years of self-observation, that it is not in hours of clearness and vigor that this doctrine (that of material atheism) commends itself to my mind ; that in the presence of stronger and healthier thought it ever dissolves and disappears, as offering no solution of the mystery in which we dwell and of which we form a part." Upon this I have to remark that the younger pupils trained in the school are beginning to say, " We need no solution except the

solution of hereditary experience ; " and some of them
will add, " We do not wish to be troubled in our em-
ployments and pleasures by any solution drawn from
a world of which we have no evidence, and which is
at best a world of darkness." I am also tempted to
say that we doubt whether it is by " self-observation " of feelings, which may vary from day to day,
and from hour to hour, that we are most likely to
get a reasonable and settled conviction. I venture
farther to hint that the theoretical opinion which
Prof. Tyndall holds, and to which he is seeking to
proselytize others, may be fostering these hours of
" weakness and doubt" of which he speaks, and hin-
dering those times of " stronger and healthier thought "
which would lead him to find a " solution of the mystery of the world in which we dwell," — not to be found,
he acknowledges, in a material atheism, but surely to
be found somewhere.

He believes in a region " outside of science," and
admits " the unquenchable claims of the emotional
nature." " Physical science cannot cover all the de-
mands of man's nature." He tells us in the Preface
to the seventh edition, " No atheistic reasoning can, I
hold, dislodge religion from the heart of man. Logic
cannot deprive us of life, and religion is life to the
religious. As an experience of consciousness, it is
perfectly beyond the assaults of logic." But is there
not a risk of this blank system undermining our
grander sentiments, by showing that this region outside of science is a region of darkness ? Our feelings,
in order to be permanent, and that they may not be

killed by the malaria of "weakness and of doubt," must be founded on *conviction*, and on a conviction which can justify itself. He who removes the ground of the conviction is doing as much as within him lies to undermine and scatter the emotions. Nature can raise within us feelings of awe, sublimity, and love, only so far as it is supposed to be pervaded by intelligence and goodness. What are these feelings, what their nature and origin, that we cherish them, or allow them to have any influence over us? What is this religion placed in the heart of man? Are they simply the product of atoms that have fortunately combined in a certain way in a time of "stronger and healthier thought," but may separate in an immediately succeeding hour of "weakness and of doubt"? If they are not, then we have here something which atoms cannot explain, and the whole theory is left in ruins. If they are, then the feelings will be cherished only when the atoms happen to meet and form them, and are in themselves no better than no feelings, or feelings of "weakness and of doubt." True, the tendencies gendered by hereditary training, or by the spirit prevailing around, may continue these nobler feelings for a time after the conviction and belief have gone; but it will be only for a time, and they will ere long die down into indifference, — just as the glow of the evening sky fades speedily into darkness, after the sun, whose beams produced it, sinks beneath the horizon. I am convinced that the tendency of this empty theory, and its actual influence, so far as it is adopted by the rising generation, is to uproot those grander sentiments of awe and of love, which are the most interesting, enlivening, and influential elements in our moral and religious nature. Will reverence and confidence, will inspiring hope and fervent affection, continue when men believe only in the interaction of atoms in a closed globe surrounded by darkness which may be felt? But Prof. Tyndall is right when he speaks of "the unquenchable claims of the emotional nature." Our natural and spontaneous feelings will be found stronger in the end than any artificial form of speculative unbelief; and they will burst forth at times like a fountain, in spite of all the efforts to repress them. But they have such power because the waters are deep down in our nature and constitution, and fed by the sky above and the earth around, penetrated by heaven-descended influences.

530 Broadway, New York,
March, 1875

Robert Carter & Brothers'

NEW BOOKS.

AUTOBIOGRAPHY AND MEMOIR OF THOMAS GUTHRIE, D.D. 2 vols. 12mo. $4.00.

" It is told in the chattiest, simplest, most unaffected way imaginable, and the pages are full of quaint, racy anecdotes, recounted in the most characteristic manner." — *London Daily News.*

THE WORKS OF THOMAS GUTHRIE, D.D.

9 vols. In a box. $13.50. (The vols. are sold separately.)

CHRISTIAN THEOLOGY FOR THE PEOPLE.

By Willis Lord, D.D., LL.D. 8vo. $4.00.

CHRISTIANITY AND SCIENCE. A Series of

Lectures, by Rev. A. P. Peabody, D.D., of Harvard College. $1.75

THE SCOTTISH PHILOSOPHY. Biographical,

Expository, Critical, from Hutcheson to Hamilton. By James McCosh, LL.D., President of Princeton College. 8vo. $4.00.

By the same Author.

The Method of Divine Government $2.50	Defence of Fundamental Truth $3.00
Typical Forms 2.50	Logic 1.50
The Intuitions of the Mind . 3.00	Christianity and Positivism 1.75

IDEAS IN NATURE, OVERLOOKED BY DR.
TYNDALL. Being an Examination of Dr. Tyndall's Belfast Address. By JAMES McCOSH, D.D , LL.D. 12mo. Paper, 25 cents; cloth, 50 cents.

THE WORKS OF JAMES HAMILTON, D.D.
Comprising : —

ROYAL PREACHER.
MOUNT OF OLIVES.
PEARL OF PARABLES.
LAMP AND LANTERN.

GREAT BIOGRAPHY.
HARP ON THE WILLOWS.
LAKE OF GALILEE.
EMBLEMS FROM EDEN.

LIFE IN EARNEST.

In 4 handsome uniform 16mo volumes. $5.00.

NATURE AND THE BIBLE. By J. W. DAWSON,
LL.D., Principal of McGill University, Montreal, Canada. With 10 full-page illustrations. $1.75.

ALL ABOUT JESUS. By the Rev. ALEXANDER
DICKSON. 12mo. $2.00.

THE SHADOWED HOME, and the Light Beyond.
By the Rev. E. H. BICKERSTETH, author of "Yesterday, To-day, and Forever." $1.50.

EARTH'S MORNING; or, Thoughts on Genesis.
By the Rev. HORATIUS BONAR, D.D. $2.00.

THE RENT VEIL. By Dr. BONAR. $1.25.

FOLLOW THE LAMB; or, Counsels to Converts.
By Dr. BONAR. 40 cents.

AN EDEN IN ENGLAND. A Tale. By A.L.O.E.
8 full-page Engravings. 16mo, $1.25; 18mo, 75 cents.

FAIRY FRISKET; or, A Peep at Insect Life. By
A.L.O.E. 75 cents.

THE LITTLE MAID AND LIVING JEWELS.
By A.L O.E. 75 cents.

THE SPANISH CAVALIER. A Tale of Seville.
By A.L.O.E.

---

* CARTERS' 50–VOLUME S. S. LIBRARY.
No. 2. Net, $20.00.

These fifty choice volumes for the Sabbath School Library, or the home circle, are printed on good paper, and very neatly bound in fine light-brown cloth. They contain an aggregate of 12,350 pages, and are put up in a wooden case. The volumes are all different from those in Carters' Cheap Library, No. 1; so that those who have No. 1, and like it, can scarcely do better than send for No. 2. *The volumes are not sold separately.* There is no discount from the price to Sabbath School Libraries.

Also, still in stock,

* CARTERS' CHEAP SABBATH-SCHOOL LI-
BRARY. **No. 1.** 50 vols. in neat cloth. In a wooden case. Net, $20.00.

TIM'S LITTLE MOTHER. By Punot. Illustrated.
$1.25.

FROGGY'S LITTLE BROTHER. By Brenda.
Illustrated. 16mo. $1.25.

FLOSS SILVERTHORN. By Agnes Giberne.
16mo. $1.25.

ELEANOR'S VISIT. By Joanna H. Mathews.
16mo. $1.25.

MABEL WALTON'S EXPERIMENT. By Joanna
H. Mathews. 16mo. $1.25.

ALICE NEVILLE, and RIVERSDALE. By C. E.
Bowen, author of "Peter's Pound and Paul's Penny." $1.25.

THE WONDER CASE. By the Rev. R. NEWTON, D.D. Containing :—

BIBLE WONDERS $1.25	LEAVES FROM TREE OF LIFE . $1.25
NATURE'S WONDERS 1.25	RILLS FROM FOUNTAIN . . . 1.25
JEWISH TABERNACLE 1.25	GIANTS AND WONDERS . . . 1.25

6 vols. In a box. $7.50.

THE JEWEL CASE. By the Same. 6 vols. In a box. $7.50.

GOLDEN APPLES ; or, Fit Words for the Young. By the Rev. EDGAR WOODS. 16mo. $1.00.

By the Author of

" THE WIDE WIDE WORLD."

THE LITTLE CAMP ON EAGLE HILL. $1.25.

WILLOW BROOK. $1.25.

SCEPTRES AND CROWNS. $1.25.

THE FLAG OF TRUCE. $1.25.

By the same Author.

THE STORY OF SMALL BEGIN-	HOUSE OF ISRAEL $1.50
NINGS. 4 vols. In a box . . $5.00	THE OLD HELMET 2.25
WALKS FROM EDEN 1.50	MELBOURNE HOUSE 2.00

By her Sister.

THE STAR OUT OF JACOB . . $1.50	HYMNS OF THE CHURCH MILI-
LITTLE JACK'S FOUR LESSONS . 0.60	TANT $1.50
STORIES OF VINEGAR HILL. 6	ELLEN MONTGOMERY'S BOOK-
vols.. . . . . 3.00	SHELF. 5 vols.. . . . . . . 5.00

A LAWYER ABROAD. By HENRY DAY, Esq. 12 full-page Illustrations. $2.00.

* 9 7 8 3 3 3 7 4 2 9 4 1 6 *